BestMasters

Springer awards „BestMasters" to the best master's theses which have been completed at renowned universities in Germany, Austria, and Switzerland.

The studies received highest marks and were recommended for publication by supervisors. They address current issues from various fields of research in natural sciences, psychology, technology, and economics.

The series addresses practitioners as well as scientists and, in particular, offers guidance for early stage researchers.

Richard Steinert

Effect of Noise on a Model Thermoacoustic System at its Stability Boundary

 Springer Spektrum

Richard Steinert
Berlin, Germany

BestMasters
ISBN 978-3-658-13822-6   ISBN 978-3-658-13823-3 (eBook)
DOI 10.1007/978-3-658-13823-3

Library of Congress Control Number: 2016936978

Springer Spektrum

Printed on acid-free paper

This Springer Spektrum imprint is published by Springer Nature
The registered company is Springer Fachmedien Wiesbaden GmbH

Acknowledgments

The research documented in this publication represents the final thesis of my studies in the master degree course of Engineering Science. It has been carried out at the Institute of Fluid Dynamics and Technical Acoustic (ISTA) at the Technische Universität Berlin.

I want to thank my supervisor Dr. Lipika Kabiraj for her great patience and believe in me. She showed me how enthusiasm and dedication brings out the best in people and work.
Further I want to thank my colleague Aditya Saurabh for the years of great work together. The experience I gained during that time was fundamental to the results presented here.
I wish both all the best on their future paths.

Richard Steinert

Contents

List of Figures

List of Tables

List of Symbols

Symbol	Description	Unit
d	inner diameter of combustor test rig	mm
d_c	diameter of circumscribing circle of perforated area	mm
d_h	diameter of flame holder holes	mm
d_{mh}	diameter of microphone connection	mm
f	frequency	Hz
f_{inst}	dominant frequency of thermoacoustic instability	Hz
f_p	peak frequency	Hz
δf_p	deviation of peak frequency	Hz
Δf	width of spectral peak at half height H	Hz
l	total length of combustor test rig	mm
l_{ds}	length of downstream duct of combustor test rig	mm
l_{us}	length of upstream duct of combustor test rig	mm
$\dot{m}_a$	air mass flow rate	kg/h
$\dot{m}_f$	fuel mass flow rate	kg/h
p'	acoustic pressure fluctuations	Pa
p'_0	acoustic pressure fluctuations at cold flow	Pa
$p'_{inst,max}$	maximum of acoustic pressure fluctuations at unstable conditions	Pa
$p'_{inst,rms}$	rms of acoustic pressure fluctuations at unstable conditions	Pa
$p'_{st,max}$	maximum of acoustic pressure fluctuations at stable conditions	Pa
$p'_{st,rms}$	rms of acoustic pressure fluctuations at unstable conditions	Pa
$\dot{q}'$	heat release rate fluctuation	W
rms	root mean square value	$-$
v	axial velocity	m/s

Symbol	Description	Unit
β	coherence factor	–
μ	generic bifurcation parameter	–
ϕ	equivalence ratio	–
ϕ_H	equivalence ratio at Hopf bifurcation	–
ϕ_{sn}	equivalence ratio at saddle-node bifurcation	–
D	noise intensity	Pa
ΔD	noise intensity step	Pa
H	height of spectral peak	arb.
NIB	noise induced bimodality	–
NIT	noise induced transition	–
Re	Reynolds number	–

1 Introduction

The combustion of natural gas in stationary gas turbines contributes more than 20 percent to today's electricity production (e.g. IEA, 2014, p. 6). Most recent developments in the exploitation of hitherto inaccessible sources of natural gas and the continuously growing demand of primary energy render this significance likely to last for the nearer future. The increasing consumption of fossil fuels, on the other hand, is jointly responsible for society's emissions of greenhouse gases (CO_2) and pollutants, such as unburnt hydrocarbons, carbon monoxide (CO) and nitric oxides (NO_x). A possibility to reduce NO_x emissions of combustion processes is to reduce the local flame temperature. In modern gas turbines this is achieved by using a lean and premixed fuel-air mixture. The excess air which does not take part in the reaction (lean conditions) cools the flame and the mixing of fuel and air prior to reaching the reaction zone (premixed) provides a homogeneous fuel distribution and thus avoids areas of higher local flame temperatures (Tacina, 1990).

While being effective in reducing NO_x emissions gas turbines using lean premixed combustor technology are prone to an unstable combustion behaviour known as thermoacoustic instability. The instability arises when the unsteady heat release of the flame interacts with the acoustic modes of the combustion chamber in a self-sustaining feedback loop. Consequences thereof are organised pressure oscillations of high amplitude which lead to increased fatigue and thus reduce the lifetime of the turbine's components (Lieuwen and Yang, 2005; Huang and Yang, 2009).

Thermoacoustic instabilities have been reported first by Rijke (1859) and are known in industrial applications such as furnaces, rocket motors and gas turbines (Merk, 1957; Barrere and Williams, 1969). Research since the 1990s has resulted in a reasonable understanding of the mechanisms leading to the onset of instabilities (e.g. Straub and Richards, 1998; Lieuwen, 1999). However, due to the complex and nonlinear behaviour of the combustion process and its interaction with the acoustic field (Dowling, 1997, 1999) a comprehensive and closed formulation of the physical phenomena has yet to be found.

Noise has been identified as a contributing factor to the complexity of thermoacoustic instabilities since the 1970s. It has been found that noisy excitations acting on thermoacoustic systems have the ability to generate coherent response in the form of acoustic oscillations (Chiu et al., 1973; Chiu and Summerfield, 1974; Strahle, 1978). Lieuwen and Banaszuk (2005) showed for a linear reduced-order model of combustion dynamics that parametric noise has the ability to render a deterministically stable combustion process unstable. Since strong noise from various sources (e.g. combustion, flow separation) is ever-present in industrial gas turbines it is an important aspect to understand as thoroughly as possible. A phenomenon that has attracted much attention in more recent works is known as triggering or hard excitation: A linearly stable system operating at the border to unstable conditions may be excited from a steady state to a limit cycle oscillation by an acoustic triggering pulse (Burnley and Culick, 2000; Lieuwen, 2002). Kabiraj and Sujith (2011) showed that the limit cycle oscillation and the steady state are separated by an unstable limit cycle which determines the necessary triggering amplitude of a single frequency excitation. Jegadeesan and Sujith (2013) demonstrated experimentally that triggering with Gaussian white noise, referred to as noise induced transition (NIT) (Horsthemke, 1984), is possible even when the triggering amplitude is much smaller compared to the necessary amplitude of the single frequency forcing.

The role of noise as a catalyst or an enhancer to structured motion has been investigated in the context of other nonlinear systems, too. For example Benzi et al. (1981, 1982) suggested the concept of stochastic resonance (SR) to be responsible for the periodic recurrence of the Earth's ice ages: A weak external periodic forcing due to variations in the Earth's orbit about the sun is enhanced by an "internal stochastic mechanism" and causes the bistable climate to change its state with the same frequency as the weak external forcing. Since then SR has also been found amongst others in electric circuits (Fauve and Heslot, 1983), laser applications (McNamara et al., 1988) and crayfish cells (Douglass et al., 1993). A thorough overview to SR can be found in Wellens et al. (2004). Coherence resonance (CR), or internal SR, is found in dynamical systems without an external forcing (Pikovsky and Kurths, 1997). In this case random noise induces a coherent response in frequencies intrinsic to the system. CR has been found as a precursor to subcritical bifurcations theoretically (Neiman et al., 1997) and experimentally in electrochemical and electrical systems (Kiss et al., 2003; Zakharova et al., 2013). To the knowledge of the author it has not been shown in thermoacoustic systems yet.

This work addresses the effects of acoustic noise on a thermoacoustic system in the vicinity of its stability boundary. The combustor that is utilised in this work uses a lean mixture of air and methane. In order to have control over the magnitude of the noisy excitation in the system an existing model combustor at ISTA has been developed further to generate minimal intrinsic noise. Variation of the fuel mass flow $\dot{m}_f$ gives control over the equivalence ratio ϕ and by that over the distance of the operating point to the stability boundary. Excitation is provided by two loudspeakers upstream of the flame imposing pressure oscillations on the fuel-air mixture; the system response is detected by means of pressure and heat release rate measurements.

The aim of this study is to provide insight to the vast field of noise induced phenomena in thermoacoustic systems by aiding in and interpreting data in the context of current research.

2 Theoretical Background

This chapter will introduce theoretical tools as well as physical concepts that are applied in this work. It is divided into two sections. The first will discuss the current understanding of thermoacoustic instabilitys and their interplay with noise. The second section is dedicated to dynamical systems theory: A mathematical approach to gain qualitative statements about the asymptotic evolution of systems described by ordinary differential equations (ODEs).

2.1 Thermoacoustic Instability

Thermoacoustic instability has been described in a scientific way first by Rijke (1859). Rijke observed it in a vertical pipe of roughly $1\,m$ height and $0.1\,m$ diameter with a gauze, located in the lower half of the duct. The gauze was heated by a flame beneath the lower end of the pipe. Due to thermal convection a mean air flow through the pipe established. When the gauze was heated sufficiently a loud tone in the frequency range of the first acoustic mode of the pipe appeared. Derivations of the so called Rijke tube are still used to study thermoacoustic instabilities due to its simple geometry and easy to control parameters (Matveev, 2003; Moeck et al., 2009; Juniper, 2010)

Rayleigh (1896) gave a first explanation to the effect: "If heat be given to the air at the moment of greatest condensation, or be taken from it at the moment of greatest rarefaction, the vibration is encouraged. On the other hand, if heat be given at the moment of greatest rarefaction, or abstracted at the moment of greatest condensation, the vibration is discouraged." In other words, if the unsteady heat release rate $\dot{q}'$ of the heat source is in phase with the pressure fluctuations p' a self-sustaining feedback loop is established. In a mathematical form, stating a necessary condition, it can be written as:

$$\int_V p'\dot{q}'dV > 0, \tag{2.1}$$

with V being the volume of the respective confinement. Since the heat transfer to the air is dominated by convection (in the case of gauze as a heat source), $\dot{q}'$ is proportional to the velocity fluctuations v' at the location of the heat source. However, due to the inertia of the transfer process a phase difference between $\dot{q}'$ and v' appears; this enables the pressure fluctuations p', which are shifted against v' by π, to be partially in phase with the heat release fluctuations, satisfying Rayleigh's criterion (equation 2.1). If the energy which is transferred from heat to acoustic energy by this process exceeds dissipation due to acoustic damping a thermoacoustic instability appears (Huang and Yang, 2009).

As stated above, Rayleigh's criterion defines a necessary condition for the onset of an instability. It does not, however, capture e.g. dissipation or energy transfer between acoustic modes and thus does not serve to predict thermoacoustic instabilities.

2.2 Dynamical Systems Theory

Dynamical systems theory is a vast field of research. The intention of this section is to present the idea of the concept and introduce the terms used later on in this work. A clear and comprehensive elaboration of the subject can be found in Strogatz (2000).

Dynamical systems theory approaches the solving of the set of n ODEs (equation 2.2), depending on the n-dimensional vector $\vec{x}$, its time derivative and a parameter μ, in a graphical way.

$$\frac{d\vec{x}}{dt} := \dot{\vec{x}} = F(\vec{x}, \mu) \tag{2.2}$$

The solution $\vec{x}(t)$ to a given set of initial values $\vec{x}(t_0)$ is viewed as a curve, the trajectory, in the n-dimensional space with the coordinates $(x_1, x_2, \ldots, x_n)$ called *phase space*. For example, the harmonic oscillator defined by equation 2.3 assigns the vector $(\dot{x}_1, \dot{x}_2) = (x_2, -\mu x_1)$ to each point (x_1, x_2) in the two dimensional phase space ($n = 2$) of the oscillator.

$$\dot{x}_1 = x_2$$
$$\dot{x}_2 = -\mu x_1$$

(2.3)

Drawing the *phase portrait*, starting at an arbitrary point $(\dot{x}_{1,0}, \dot{x}_{2,0})$, following the vectors defined by 2.3 one will see that the trajectories are circles around the origin. The only exception being the point $(\dot{x}_{1,0}, \dot{x}_{2,0}) = (0, 0)$. The trajectory starting here will remain at the origin of the phase space. By assigning the (angular) position θ to x_1 and the velocity v to $x_2 = \dot{x}_1$ the face portrait gives an intuitive picture of the harmonic oscillator without actually (analytically) solving the ODE: The system will either oscillate at an amplitude defined by the initial conditions (and a frequency defined by $\sqrt{\mu}$), or remain stationary at the origin if no initial displacement is imposed (2.1.a). These two states are considered *stable* in dynamical systems parlance: Without external forcing, the system will not leave the respective state. In particular the oscillatory state is referred to as a stable *limit cycle*, the origin as a stable *focus* or *fixed point*.

The introduced phase portrait of 2.3 will not change its general appearance when μ changes. However, for the undamped pendulum in a gravitational field, for example, a third stable state besides oscillating around the resting position and remaining still is possible: The continuous rotation around the pendulum's bearing. Figure 2.1.b shows the rotating state as wavelike trajectories with either all positive or all negative velocity v.

The transition from oscillation to rotation can be achieved by either choosing a sufficiently high initial velocity v_0 or by reducing the gravitational acceleration g. The latter is of importance, despite remaining a theoretical example, because the transition from one state to the other happened by changing the system parameter, in this case g. This could be done while the pendulum oscillates; such a change in a system's qualitative behaviour is known as *bifurcation*, the parameter g plays the role of the *bifurcation parameter*. A bifurcation will come along with qualitative changes in the system's phase portrait: New stable states may emerge and old ones disappear. In a *bifurcation diagram* the bifurcation parameter μ is plotted on the x-axis and a characteristic state variable x on the y-axis. The stable states of x as a function of μ are represented as lines in the diagram and by that the dynamics of x are visualised.

a)   b)

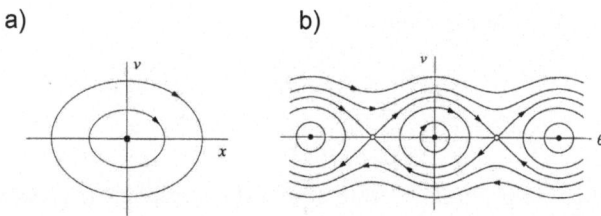

Figure 2.1: Phase portraits of the harmonic oscillator (a) and the undamped pendulum in a potential field (b) (Strogatz, 2000).

The definition of stability given in this section can be contradicting to the terminology used for thermoacoustic systems. When a thermoacoustic instability appears, the system is still in a stable state from the dynamical systems point of view, namely a stable limit cycle. What has become unstable during transition, however, is the fixed point at the origin of the phase plane; disturbances of any amplitude to the fixed point will lead to the limit cycle oscillation referred to as thermoacoustic instability.

3 Experimental Setup

This chapter addresses the burner test rig and the measuring equipment that has been used for the experiments reported in this work. The burner used here is a modification of the burner used by Kabiraj et al. (2013). All experiments were carried out at the energy lab at ISTA[1], TU Berlin.
An overview of the main characteristics of the test rig is given in table 3.1.

height	l	$1,185\,mm$
length of upstream section	l_{us}	$570\,mm$
length of downstream section	l_{ds}	$615\,mm$
inner diameter	d	$105\,mm$
reactant		atmospheric oxygen
air mass flow rate	$\dot{m}_a$	$1.30\,kg/h$
fuel		methane
fuel mass flow rate	$\dot{m}_f$	$52.5 - 54.0\,g/h$
equivalence ratio	ϕ	$0.695 - 0.715$
Reynolds number	Re	147

Table 3.1: Operating conditions and characteristics of the test rig. The Reynolds number is calculated with respect to the diameter of the holes in the flame holder plate and the mean velocity at that point.

3.1 Burner Test Rig

The burner consists of an upstream and a downstream duct with an inner diameter of $d = 105\,mm$ and a total length of $l = 1,185\,mm$, the upstream duct has a length of $l_{us} = 570\,mm$, the downstream duct has a length of $l_{ds} = 615\,mm$; both ducts are aligned vertically. The upstream duct is closed at the bottom

---

[1]Institut für Strömungsmechanik und Technische Akustik (ISTA), Fachgebiet Experimentelle Strömungsmechanik, Prof. Dr.-Ing. C. O. Paschereit

except for an inlet valve for the fuel-air mixture. A silencing device consisting of three alternating layers of aluminium discs and fire proof foam reduces the noise due to the entering fuel-air mixture to $p'_0 = 0.18\,Pa\,(rms)$. Two loudspeakers are attached to the side of the upstream duct directly downstream of the silencer; their input signal is generated by a DS1103 PPC control board. The downstream duct has a fully open outlet. To make the flame optically accessible, the lower half of the downstream duct (300 mm) is made of quartz glass. A schematic illustration of the setup is given in figure 3.1.

The upstream and downstream ducts are separated by a perforated brass plate which acts as a flame holder. It has 91 holes with a diameter of $d_h = 2\,mm$ which are arranged in a hexagonal shape with a circumscribing circle with a diameter of $d_c = 50\,mm$ (see figure 3.2). The flow passes the holes with a mean velocity of $v = 1.13\,m/s$, resulting in a Reynolds number of $Re = 147$ and thus a laminar flow. The flame which stabilises on top of this plate creates a layer of conical flames. The design of the plate has been modified through two prior iterations to create a flame featuring a subcritical bifurcation under lean conditions.

3.2 Operating Conditions

The operating point of the system is determined by the mass flow rates of air $\dot{m}_a$ and fuel (methane) $\dot{m}_f$, implying a certain equivalence ratio ϕ. Throughout the experimental work $\dot{m}_a$ has been held constant at $\dot{m}_a = 1.3\,kg/h$. In order to set a desired equivalence ratio the fuel mass flow rate was adjusted. This procedure gives the advantage of a nearly constant total mass flow due to the small content of about four percent of fuel in the mixture.

Both flowrates, $\dot{m}_a$ and $\dot{m}_f$, are measured by coriolis flow meters and regulated by PID controlled valves. An Endress + Hauser PROMASS A80 is used for air; a Bronkhorst mini CORI-FLOW integrates both flow meter and valve for fuel. Air is measured with a nominal uncertainty of $\Delta\dot{m}_a = \pm 0.5\,\%$, fuel is measured with a nominal uncertainty of $\Delta\dot{m}_f = \pm(0.2\,\% + 6\,g/h)$ at the respective mass flow rates. The relatively low accuracy in the fuel mass flow rate measurement stems from its low absolute set point around $\dot{m}_f = 0.055\,kg/h$. How this error impacts the outcome of the experiment is discussed in chapter 4.

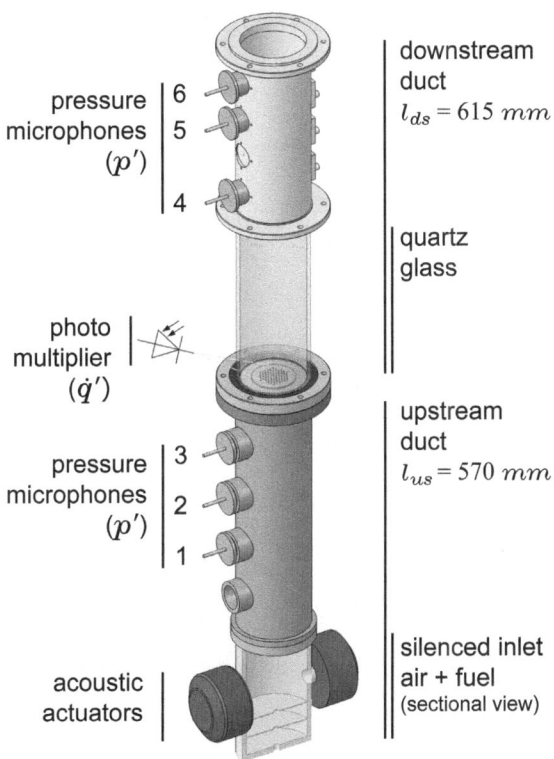

Figure 3.1: Isometric view of the burner test rig (scale 1:10).

3.3 Data Acquisition

The two main quantities which are captured during the experiments are the acoustic pressure fluctuation p' along the length of the ducts and the heat release rate fluctuation $\dot{q}'$ from the flame. The former is realised by six cooled pressure microphones (G.R.A.S. 40BP pressure microphones and G.R.A.S. 26AC preamplifier) which are mounted normal to the duct surface and which are spatially connected to the inside of the duct through a hole of $d_{mh} = 1\,mm$ diameter. Three microphones are located at the upstream duct and three at the downstream duct; their location with respect to the flame holder is given in table 3.2. The

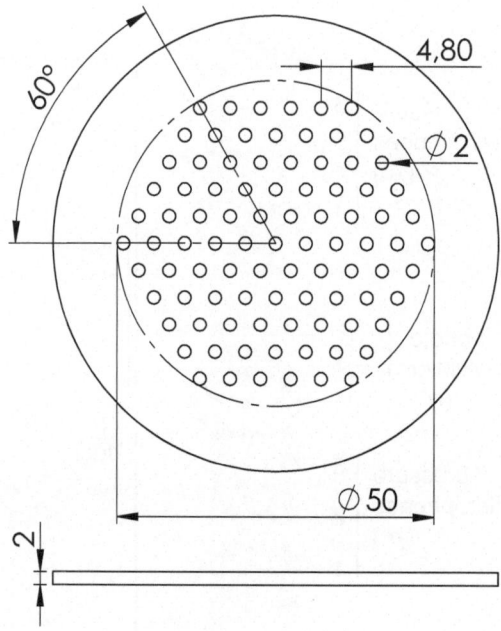

Figure 3.2: Dimensions of the flame holder plate (scale 1:1), given in mm.

microphones are calibrated to account for phase and magnitude. Calibration is done for the upstream microphones on a separated calibration tube of the same wall thickness as the duct, for the downstream microphones it is done directly at the duct by mounting the reference microphone flush to the inner surface of the duct, opposing the respective microphone. The calibrated microphones allow for a quantitative measurement of the acoustic field in the rig. Although all six microphone signals have been captured, pressure data analysed in chapter 4 originates from microphone 4, see figure 3.1. This selection is based on the calibration of the microphones; microphone 4 featured the smallest deviation from the reference microphone.

The heat release rate fluctuation $\dot{q}'$ is measured by using a photomultiplier tube directed at the flame through the quartz glass section. The incident light is filtered by a $307\,nm$ filter to only capture light that is emitted by excited CH* radicals in the flame. The light intensity at this wavelength acts as a relative measure of the flames heat release rate (Haber, 2000). The photomultiplier tube captures the integrated light emission from the entire flame and thus, provides a scalar measurement of $\dot{q}'$ in time. It is used in this work to gain qualitative statements about the heat release rate fluctuations of the flame.

The analogue signals of the pressure microphones and the photomultiplier tube are amplified by analogue amplifiers. The signals are then read by an analogue-to-digital converter (National Instruments NI cDAQ 9172 with a NI 9215 D/A module, 8192 samples/s) and stored on a PC.

microphone	position
mic. 1	$-281\,mm$
mic. 2	$-191\,mm$
mic. 3	$-101\,mm$
mic. 4	$349\,mm$
mic. 5	$479\,mm$
mic. 6	$544\,mm$

Table 3.2: Locations of pressure microphones with respect to the flame holder. Positive values point in direction of the flow (upwards). Data acquired with microphone 4 is mainly used in chapter 4.

4 Results and Discussion

This chapter will analyse and discuss the data acquired during the preceding experimental work. It is divided into three sections. The first focuses on the system behaviour without external forcing and will introduce the system's regions and boundaries of stability. In the second section, the effect of single frequency forcing on the system at bistable conditions is addressed. The central part of this work will be discussed in the third section: The effect of noise on the system in stable conditions close to the bistable region.

4.1 Determination of the System's Stability Characteristics

Time series of the acoustic pressure fluctuations p' of the flow field and the heat release rate fluctuations $\dot{q}'$ of the flame are the basis of this analysis. At stable conditions, both p' and $\dot{q}'$ were recorded for the duration of 32 seconds. Subsequently the equivalence ratio has been increased stepwise and new recordings were taken for each step. When the system became unstable, the equivalence ratio was decreased again and recordings were taken until the system re-entered stable conditions. During measurements a real-time frequency spectrum indicated the occurrence of an instability by a moderate rising amplitude across the whole spectrum and the appearance of distinct peaks. It could also be perceived as a loud and low sound from the combustor.

Figure 4.1 shows exemplary time series of pressure and heat release rate fluctuations for stable and unstable conditions. It can be seen that both state variables experience a significant amplification during instability, however, p' much stronger than $\dot{q}'$. The transitions between the states are captured in figure 4.2.a and 4.2.b; the signals grow and decay exponentially.

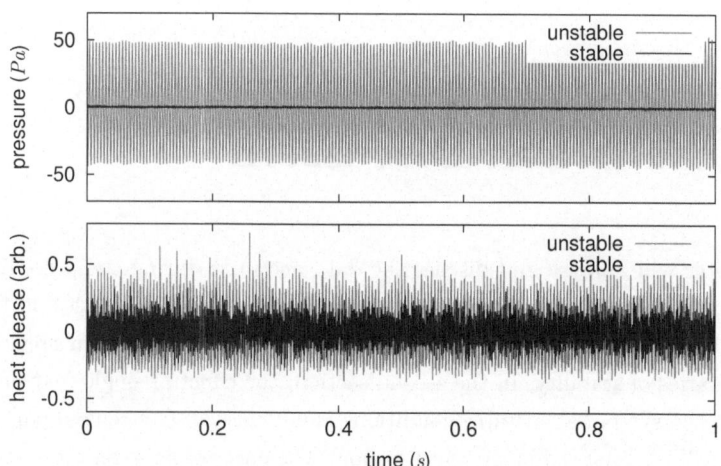

Figure 4.1: Exemplary time series of p' and $\dot{q}'$ for stable ($\phi = 0.708$) and unstable conditions ($\phi = 0.728$).

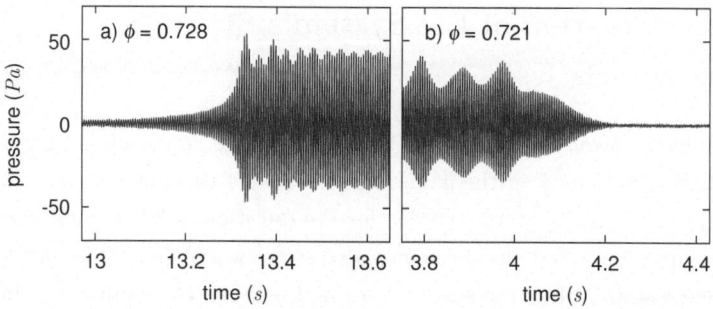

Figure 4.2: Transition between stable and self-exited state.

The frequency spectra (figure 4.3) reveal that the unstable system settles to a limit cycle oscillation, the dominating frequency of both pressure and heat release rate fluctuations during instability is identified as $f_{inst} = 192\,Hz$. Super harmonics are also present in the p' and $\dot{q}'$ spectra but are smaller in magnitude by a factor of 10^3 and 10^2, respectively. Two peaks in the stable and unstable pressure spectra at $f = 345\,Hz$ and $f = 690\,Hz$ seem not to be related to thermoacoustic coupling since they are not reflected in the heat release rate spectrum and do not change significantly between stable and unstable conditions.

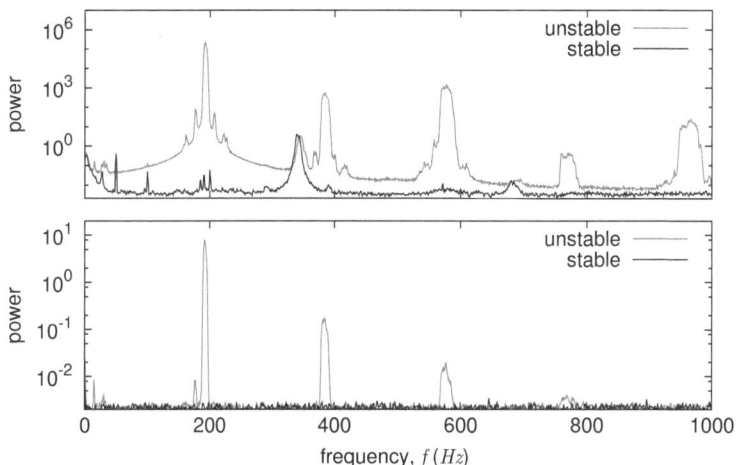

Figure 4.3: Exemplary frequency spectra of p' and $\dot{q}'$ for stable ($\phi = 0.708$) and unstable conditions ($\phi = 0.728$).

A bifurcation diagram is derived from the data collected in the above described manner. To this purpose the root-mean-square (rms) values of the pressure time series are plotted against the control parameter, ϕ. The plot illustrates the two possible states the system exhibits: The stable state close at zero amplitude and the self-excited limit cycle oscillation during thermoacoustic instability at around $p'_{inst,rms} = 26.8\,Pa$. These two states are distinct; stable intermediate states do not exist, which can be seen in the transition plots (figure 4.2). It also shows that whether the system is stable or unstable does not only depend on the equivalence ratio: In the range of $0.715 < \phi < 0.728$ the state of the system is determined by the path on which it approaches this region. The system shows hysteresis, an effect that is typical to thermoacoustic systems that undergo subcritical bifurcations (Straub and Richards, 1998; Lieuwen, 2002; Moeck et al., 2008).

The observed behaviour of the model combustor is known in dynamical systems theory as a saddle-node bifurcation and subcritical Hopf bifurcation (Strogatz, 2000). Figure 4.5 illustrates the evolution of the corresponding normal form in its two dimensional phase space. It is characterised by a stable fixed point at the origin and a stable and an unstable limit cycle. As the system parameter (the

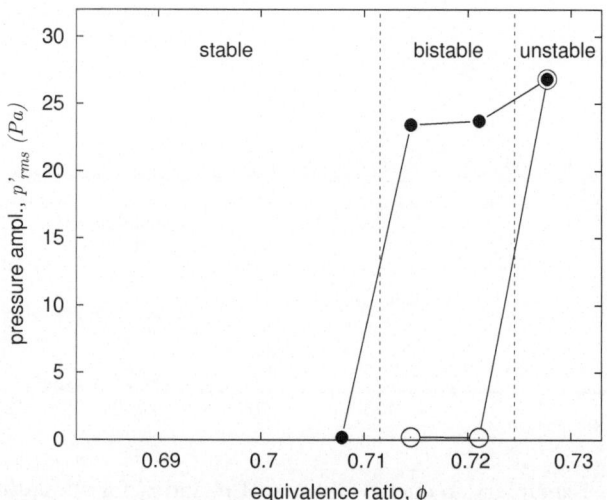

Figure 4.4: Bifurcation diagram of the unforced system. Dotted lines define stability regions. Open circles (◦) indicating experiments with increasing equivalence ratio, filled circles (•) indicating experiments with decreasing equivalence ratio.

equivalence ratio ϕ in this work, μ (arbitrary) in the figure) varies the unstable limit cycle changes its amplitude (represented in the phase portrait by its radius) and will eventually coincide with either of the stable states. At that point the two colliding states annihilate and the system undergoes a bifurcation: The state variables will suddenly move to the remaining stable state. In figure 4.5 on the right side, only the stable limit cycle remains, from all points in the phases space the system will evolve towards that attractor. This case represents the onset of the thermoacoustic instability, it is referred to as the subcritical Hopf bifurcation (at $\phi = \phi_H$). The other case (when the fixed point remains) is known as a saddle-node bifurcation (at $\phi = \phi_{sn}$), it represents the re-stabilization of the investigated combustor. The coexistence of two stable states for values $\phi_{sn} < \phi < \phi_H$ leads to the observed hysteresis.

Thermoacoustic systems have been reported to exhibit this behaviour previously in the literature: Lieuwen (2002) demonstrated it with the inlet velocity to a premixed combustor as the system parameter, Jegadeesan and Sujith (2013) and

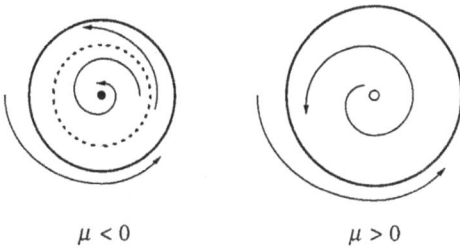

$\mu < 0$ $\mu > 0$

Figure 4.5: Phase portrait of a generic subcritical Hopf bifurcation (Strogatz, 2000). Left: Coexistence of a fixed point at the origin and a stable limit cycle (closed circle) separated by an unstable limit cycle (dashed circle). Right: After bifurcation only the stable limit cycle remains.

Kabiraj and Sujith (2012) with the location of a Bunsen type burner with respect to its confinement. Kabiraj and Sujith (2011) demonstrated experimentally the changing amplitude of the unstable limit cycle in between the two bifurcation points. Juniper (2010) and Subramanian et al. (2013) were able to estimate analytically the regions of stability and the amplitude of the unstable limit cycle for a Rijke tube model.

In the previously shown pressure time series (figures 4.1 and 4.2) of the self-excited system a low frequent modulation of the instability amplitude is observed. It also manifests in the heat release rate as shown in figure 4.6 and is depicted in the spectra at frequencies below $50\,Hz$ (figure 4.3). This behaviour points to the complex nonlinear phenomena that determine and influence the amplitude of thermoacoustic instabilities which are subject to recent publications (Noiray et al., 2011; Noiray and Schuermans, 2012).

According to the different states which the investigated thermoacoustic system can exhibit, the range of equivalence ratios is subdivided into three parts. The range $\phi < \phi_{sn}$ is considered the *stable region*: It only allows for the system do remain in an unexcited (stable) state. Between ϕ_{sn} and ϕ_H, the range where hysteresis is present, the system is either unexcited (stable) or in the self-excited limit cycle (in case of thermoacoustic instability) and thus is referred to as the *bistable region*. The region $\phi_H < \phi$ is denominated *unstable region* because the system will leave the (stable) fixed point and evolve to the self-excited limit cycle.

A summarisation of characterising quantities of the unforced system is listed in table 4.1.

pressure maximum, stable	$p'_{st,max}$	$0.65\,Pa$
pressure maximum, unstable	$p'_{inst,max}$	$52.10\,Pa$
pressure rms, stable	$p'_{st,rms}$	$0.18\,Pa$
pressure rms, unstable	$p'_{inst,rms}$	$26.81\,Pa$
dominant frequency, unstable	f_{inst}	$192\,Hz$
Hopf point	ϕ_H	0.728
saddle-node point	ϕ_{sn}	0.715

Table 4.1: Characteristics of the unforced system.

4.2 Single Frequency Excitation in the Bistable Region

In the bistable region at an equivalence ratio of $\phi = 0.721$ the flow field has been exposed to external acoustic forcing by a single frequency signal of varying amplitude using the actuators described in chapter 3. The forcing frequency has been chosen equal to the dominant frequency of the unstable limit cycle, $f_{inst} = 192\,Hz$; the duration of the signal has been set to $50\,ms$. The forcing amplitude has been increased incrementally, starting at $1.4\,Pa\,(rms)$.

When forcing with an amplitude of $2.8\,Pa\,(rms)$ the system pressure evolved to the self-sustained oscillatory state. This phenomenon called *triggering* or *hard excitation* is typical for the subcritical Hopf bifurcation. It has been shown experimentally amongst others for rocket motors (Wicker et al., 1996) and premixed model combustors (Lieuwen, 2002; Moeck et al., 2008; Jegadeesan and Sujith, 2013). Kabiraj and Sujith (2011) demonstrated that the minimal triggering amplitude is equal to the amplitude of an unstable limit cycle that exists in the bistable region (figure 4.5). It is dividing the basins of attraction of the fixed point at zero amplitude and the limit cycle oscillation. The unstable limit cycle amplitude and thus the necessary triggering amplitude depend on the equivalence ratio. In agreement with the theoretical formulation of the subcritical Hopf bifurcation (Strogatz, 2000) both limit cycles (stable and unstable) appear

at the saddle-node, having the same amplitude initially. The closer the operating
point moves towards the Hopf bifurcation, the smaller the amplitude of the
unstable limit cycle becomes. As shown in figure 4.6 (1st and 2nd column), not
every triggering pulse of $2.8\,Pa\,(rms)$ amplitude resulted in the self-sustained
oscillation. This can be explained by the small width of the bistable region, which
leads to a strong variation of the necessary triggering amplitude at relatively
small equivalence ratio fluctuations. The inherent uncertainty of the fuel mass
flow rate could be responsible for these fluctuations.

The 3rd column in figure 4.6 shows a successful attempt of triggering instability.
Like in previously shown pressure time series (figures 4.1 and 4.2) the modulated
amplitude of the instability is obvious.

Due to difficulties to repeat the triggering experiments in a systematic way a
discussion beyond a phenomenological description is not done at this point.

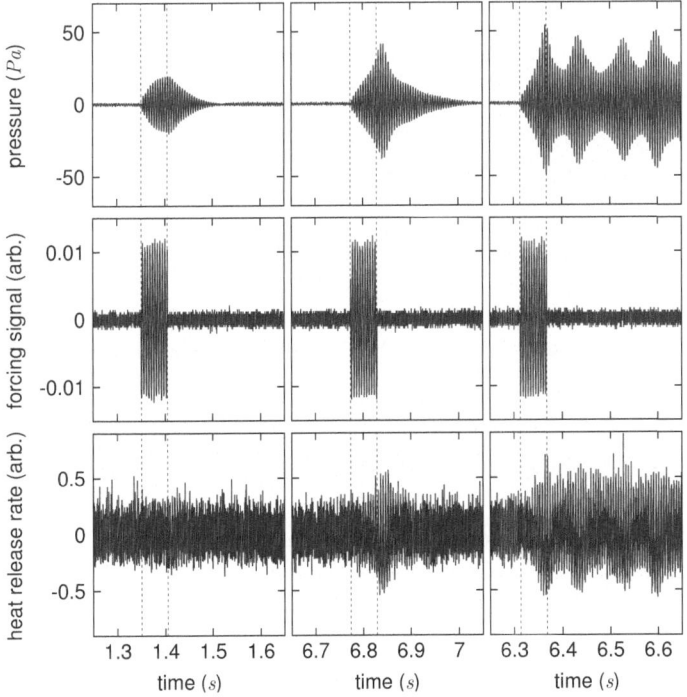

Figure 4.6: Triggering the linearly stable system in the bistable region.

4.3 Noise Induced Response in the Stable Region

The following section discusses the system's behaviour under the influence of external acoustic noise. Here the equivalence ratio has been set to stable conditions close to the bistable region ($0.695 \leq \phi \leq 0.715$). As the bifurcation diagram (figure 4.4) indicates the only stable state of the unforced system under these conditions is the unexcited state around zero amplitude. Spontaneous self-excitation and triggering are not expected in this region.

The experiments were conducted according to the following procedure: Starting at an equivalence ratio of $\phi = 0.695$ the noise intensity has been increased subsequently from $D = 1.4\,Pa$ to $D = 21.2\,Pa$ in steps of $\Delta D = 1.4\,Pa$. About 30 seconds after the noise intensity had been increased, measurements were taken for 32 seconds with a sampling frequency of 8192 samples per second. After reaching $D = 21.2\,Pa$ measurements were repeated for $\phi = 0.701$, $\phi = 0.708$ and $\phi = 0.715$. This process has then been repeated seven times for each equivalence ratio.

4.3.1 Analyses in the Time Domain

Figures 4.7 and 4.8 display the rms values of pressure and heat release rate fluctuations, p' and $\dot{q}'$, respectively as a function of the equivalence ratio ϕ. Each data point represents the average over the rms values of seven independent measurements for the same conditions. Each curve represents measurements for a given noise intensity D. The stable branch of the hysteresis loop (figure 4.4) is shown for reference. As emphasised by the dotted lines all experiments were conducted at stable conditions or at the border to the bistable region. For clarity only three selected noise intensities are plotted.

Both figures show a similar qualitative behaviour: The response of p' and $\dot{q}'$ to the excitation generally grows with the increase of the noise intensity on one hand and with the increase of the equivalence ratio on the other hand. However, at the lowest intensity ($D = 5.7\,Pa$) the acoustic pressure in the duct does not depend on ϕ until it reaches the bistable region. The heat release rate fluctuations behave similarly with a slight response to the equivalence ratio increase, as is expected for an increasing fuel flow rate. Beyond $D = 5.7\,Pa$ p' responds with amplitudes

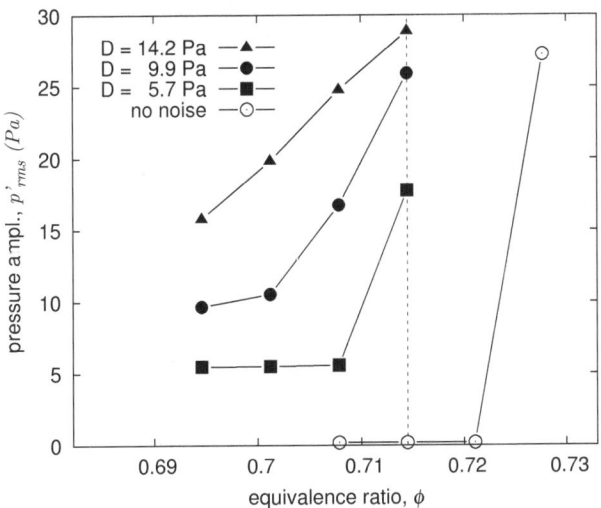

Figure 4.7: Pressure response to noisy excitation. Dotted line separates stable and bistable region.

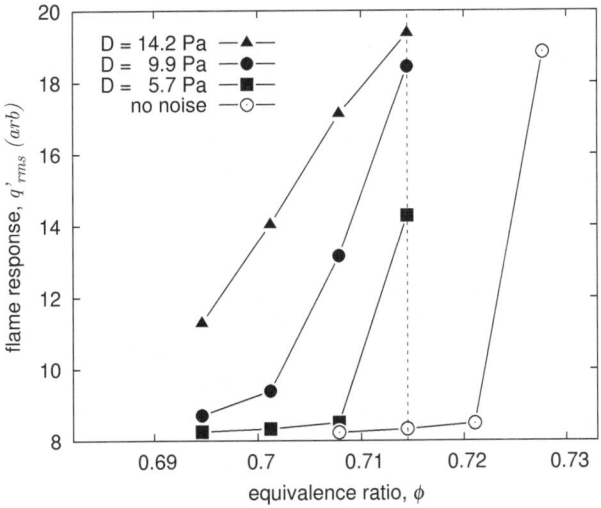

Figure 4.8: Flame response to noisy excitation. Dotted line separates stable and bistable region.

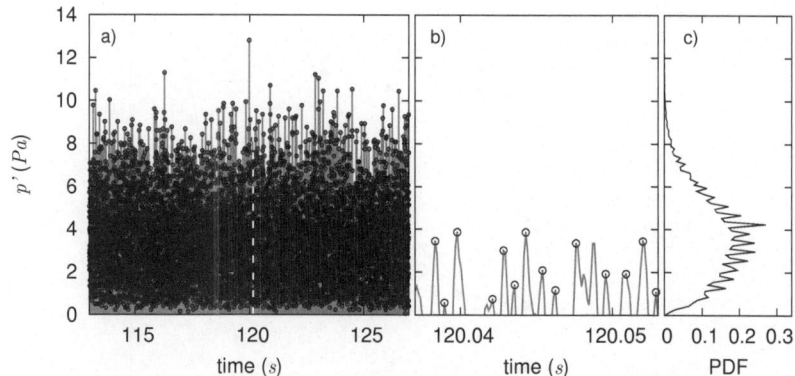

Figure 4.9: a) Pressure time series (lines) and associated amplitudes (o), dashed line indicates position of b). Only positive values are shown. b) Detail view of figure a). c) Normalised PDF derived from pressure time series.

greater than the excitation, reaching the instability amplitude of $p'_{inst} = 26.8\,Pa$ at an excitation of $D = 14.2\,Pa$ and $\phi = 0.708$.

A more detailed insight into the structure of the system response can be gained by calculating the probability density functions (PDFs) of amplitude distributions of p' and $\dot{q}'$ during the excitation. For noisy signals such as the ones looked at in this work it is not trivial to define an *amplitude*. In this work the amplitude is considered to be the maximum absolute value of the signal between two zero crossings. Figure 4.9 illustrates the method that was applied to derive the PDFs from the time series. Each PDF represents the accumulated amplitude distribution of seven independent measurements at the same conditions. To account for different frequencies of the detected amplitudes, each amplitude is weighted with the time difference to the previous detected amplitude. The amplitudes are sorted into 100 bins equally distributed from zero to the maximum value of the respective time series; the PDF's area is normalised.

Figures 4.11 to 4.14 display the PDFs for the stable equivalence ratios $\phi = 0.695$ to $\phi = 0.715$ for both p' and $\dot{q}'$ at noise intensities D between $2.8\,Pa$ and $21.2\,Pa$; figure 4.10 shows the PDFs for the non-reacting flow for comparison.

It can be seen that at $\phi = 0.695$ the PDFs for p' have a single maximum that shifts towards higher amplitudes and the curves become broader as the noise intensity

D increases. The peak amplitude is of the value as the noise intensity itself which agrees with the plots of the rms values in figures 4.7 and 4.8. This behaviour is expected as the system operates in stable conditions and the pressure field in the duct is mostly determined by the added acoustic noise. However, comparison with the non-reactive case reveals that the flame leads to a slightly broader amplitude distribution with higher maximum amplitudes when D is high; this points to the onset of the coupling of pressure field and flame. Data of the heat release rate $\dot{q}'$ supports this impression: Figure 4.11.c shows that the PDFs at noise intensities $D < 9.9\,Pa$ are not affected by the acoustic excitation. Beyond that the PDFs become broader and at $D = 18.4\,Pa$ a hint to the appearance of a second peak around 0.3 is visible.

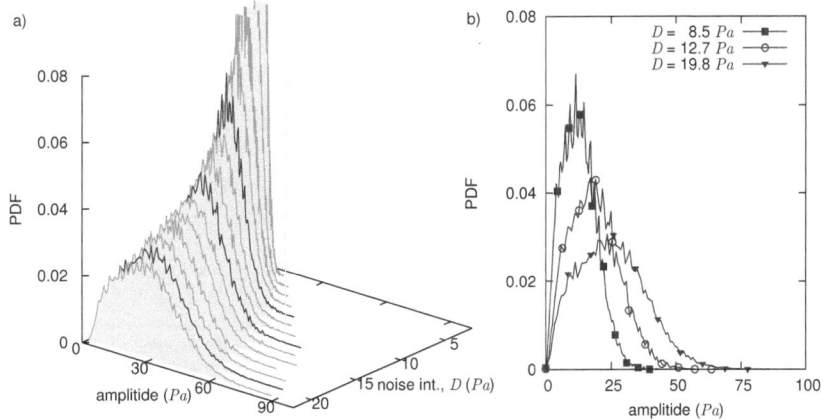

Figure 4.10: PDFs of pressure time series of non-reactive flow for noise intensities from $2.8\,Pa$ to $21.2\,Pa$. Selected PDFs are marked dark (a) and plotted separately (b).

At $\phi = 0.701$ a similar evolution of the PDFs with increasing noise intensity can be observed (figure 4.12). In this case however, the amplitude distribution of p' does not only become broader for high noise intensities but also develops a second maximum around $40\,Pa$ which even dominates the low amplitude peak around $10\,Pa$. This bimodal distribution is most pronounced at $D = 18.4\,Pa$ and becomes blurred again for higher noise intensities. The amplitude distribution for $\dot{q}'$ also clearly shows the second maximum at $D = 18.4\,Pa$. However, in contrast

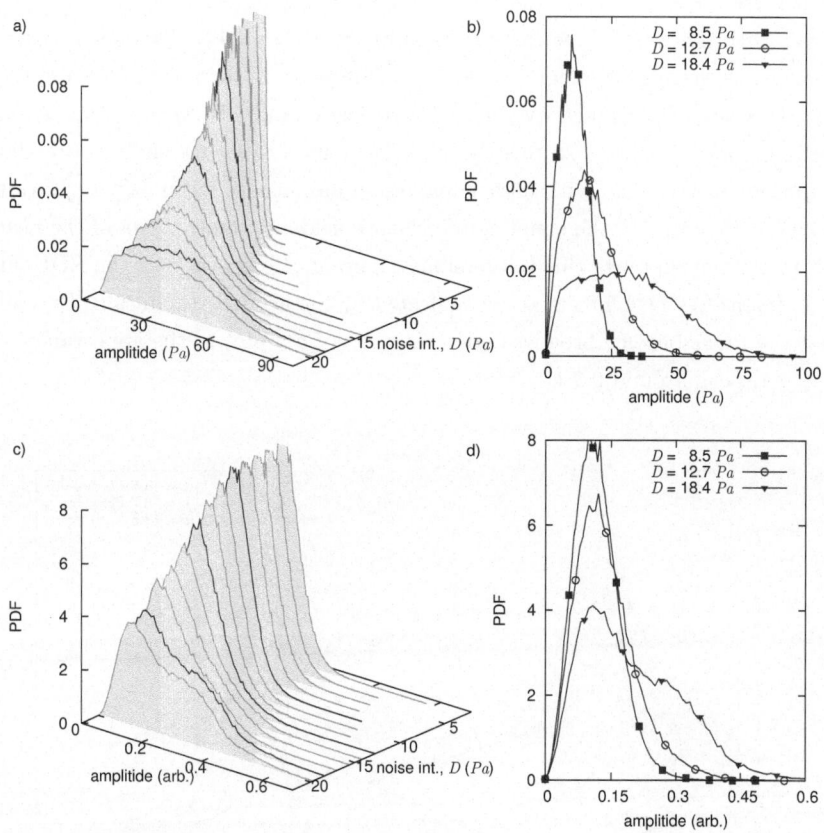

Figure 4.11: PDFs at $\phi = 0.695$ for p' (a, b) and $\dot{q}'$ (c, d). Selected PDFs are marked (a and c) and plotted separately (b and d).

to the pressure data, noise intensities beyond $D = 18.4\,Pa$ seem not to change the bimodal distribution significantly.

At $\phi = 0.708$ the trend seen at lower equivalence ratios continues (figure 4.13): The pressure field gradually becomes broader with increasing noise and develops a bimodal distribution which is most pronounced at a given noise intensity, in this case at $D = 12.7\,Pa$. Beyond that the distribution becomes broader. The location of the peak amplitudes shifts to higher values with increasing noise. On the other hand, the heat release rate is only marginally affected by low

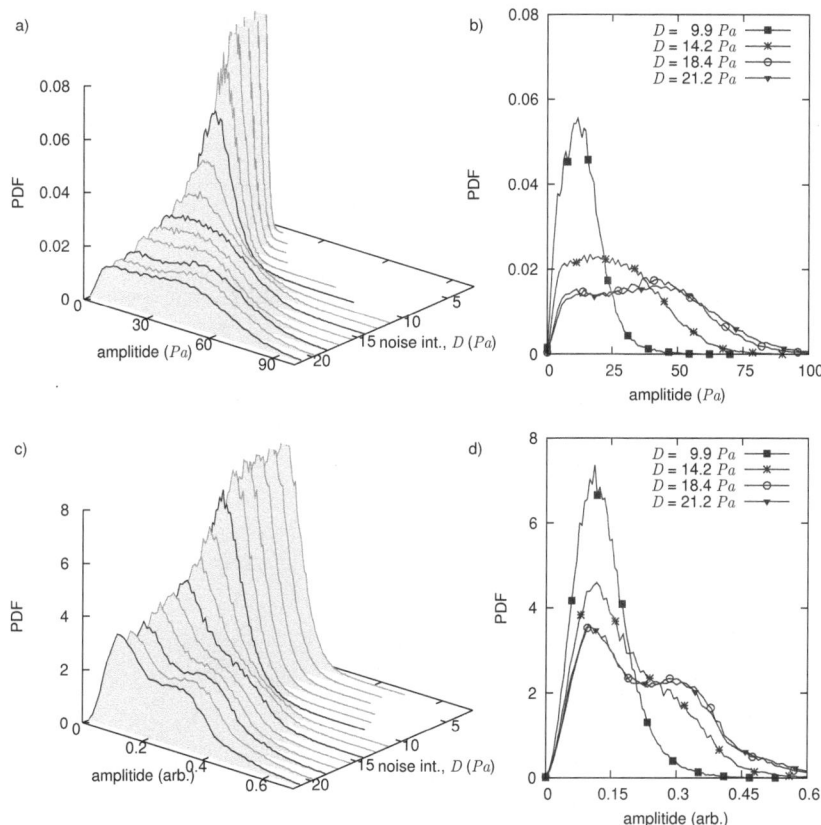

Figure 4.12: PDFs at $\phi = 0.701$ for p' (a, b) and $\dot{q}'$ (c, d). Selected PDFs are marked (a and c) and plotted separately (b and d).

excitation amplitudes. A transition to a bimodal distribution can be observed at intermediate noise intensities, a further increase in D affects the distribution only marginally. The locations of the peak amplitudes of the heat release rate fluctuations do not shift with increasing noise.

In contrast to the lower equivalence ratios the pressure PDFs in this case show a sudden jump from the single maximum distribution to the bimodal distribution when the noise intensity increases from $D = 11.3\,Pa$ to $D = 12.7\,Pa$.

Figure 4.14 shows the PDFs for $\phi = 0.715$. Here, at the border to the bistable

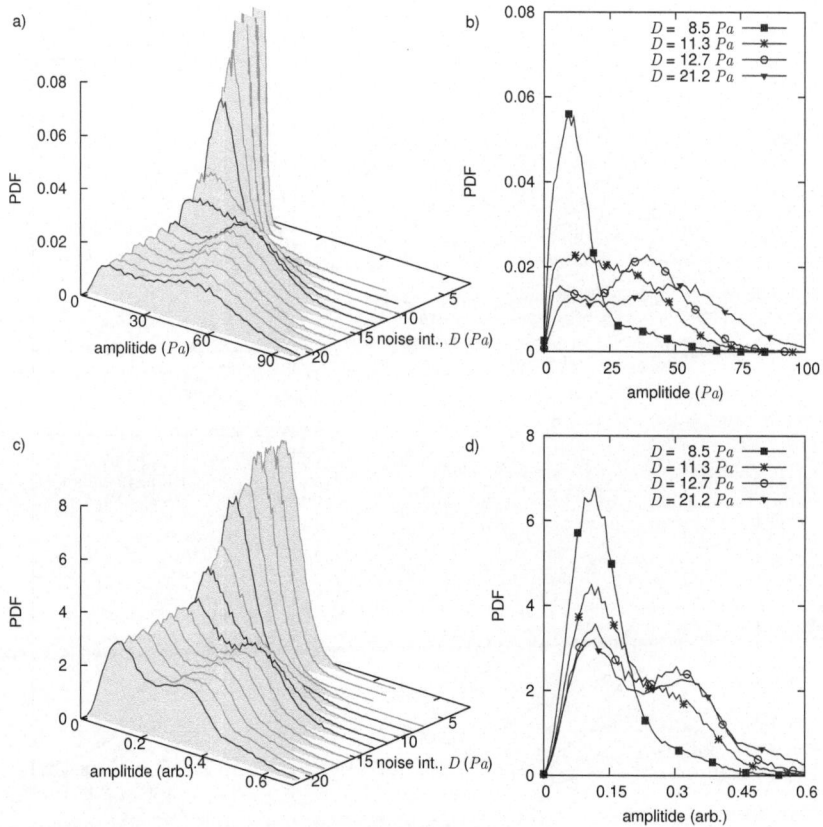

Figure 4.13: PDFs at $\phi = 0.708$ for p' (a, b) and $\dot{q}'$ (c, d). Selected PDFs are marked (a and c) and plotted separately (b and d).

region the smallest noise intensity $D = 2.8\,Pa$ already provokes a bimodal response that can be seen in both p' and $\dot{q}'$; the maxima are more distinct as they were for lower ϕ-values with the probability density near zero in between the two peaks.

As it has been seen before the pressure amplitude distribution is heavily affected by noise across the whole range of noise intensities: Higher intensities lead to the suppression of the maxima in favour of a broader distribution and to the shift of the maxima towards higher amplitudes. In contrast, the PDFs of $\dot{q}'$ are subject

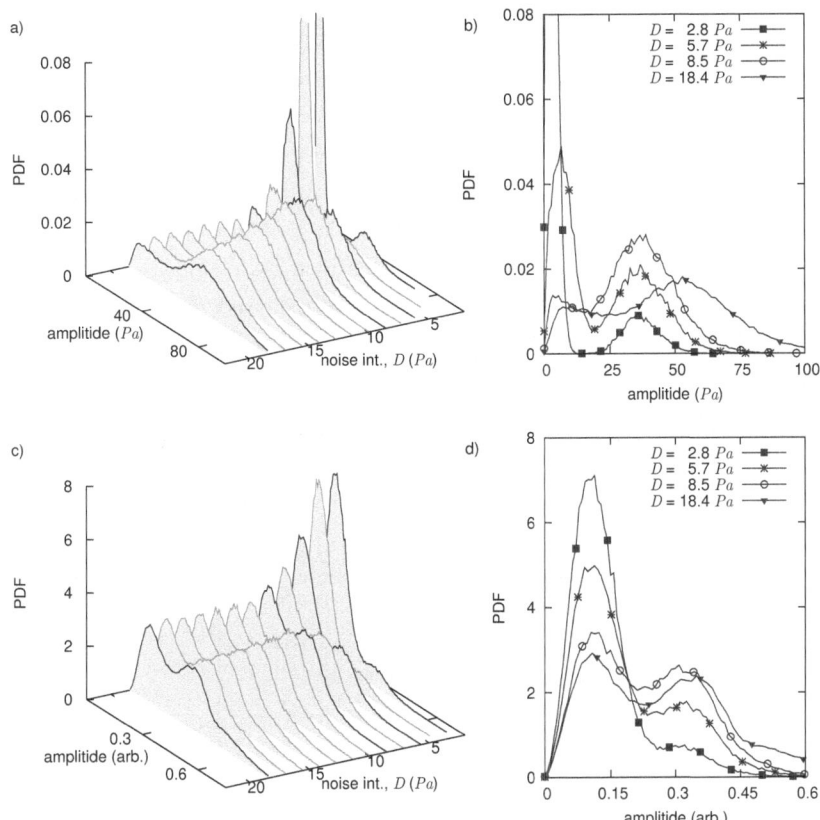

Figure 4.14: PDFs at $\phi = 0.715$ for p' (a, b) and $\dot{q}'$ (c, d). Selected PDFs are marked (a and c) and plotted separately (b and d).

to only minor changes after the bimodal form has developed. Nevertheless, a broadening effect at high noise intensities can be seen for $\dot{q}'$ in figure 4.14.d.

The PDFs support the main result of the rms plots: Although operating in stable conditions the system is susceptible to the acoustic excitation with Gaussian white noise; with growing excitation intensity and closer to the bistable region the acoustic pressure oscillations and the heat release rate fluctuations become more prominent. What the PDFs reveal is that depending on the noise intensity the amplitudes of both p' and $\dot{q}'$ exhibit a bimodal distribution: Two oscillation

amplitudes show a maximum in the probability distribution. This phenomenon
will be referred to as *noise induced bimodality* (NIB).

The bimodal pressure amplitude distribution is most pronounced at intermediate
noise intensities. At low intensities pressure oscillates around zero amplitude; at
high intensities the pressure PDFs become broader and the high amplitude maximum shifts to higher values. Zakharova et al. (2010) showed similar numerical
and analytical results for the Duffing–Van der Pol oscillator driven by Gaussian
white noise at stable conditions close to the bistable region, see figure 4.15.
A successive publication (Zakharova et al., 2013) showed the same behaviour
experimentally for the electronic realisation of said oscillator. Zakharova et al.
(2010, 2013) associate the change in the PDFs between unimodal to bimodal
with stochastic phenomenological bifurcations (P-bifurcations (Arnold, 2013)).

It is worth pointing out that the term *bimodal* in the context of amplitude

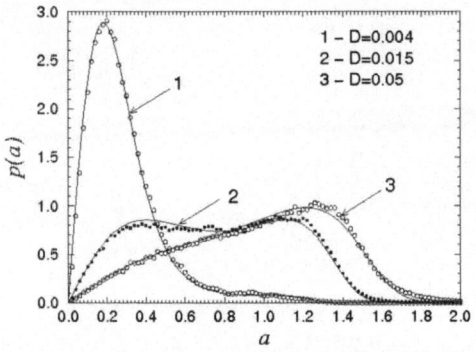

Figure 4.15: PDF for the Duffing-Van der Pol oscillator driven by Gaussian
white noise (Zakharova et al., 2010). Numerical (points) and analytical results
(lines).

probability distributions in other publications (e.g. Lieuwen, 2002; Noiray and
Schuermans, 2012) relates to the two maxima at the negative and the positive
value of the amplitude of an oscillation. By the definition in this work these
would be considered unimodal distributions.

What can be seen further in the PDFs is the difference in the evolution of the p'-
and $\dot{q}'$-curves with increasing noise level. From a certain level on the amplitude
distributions of the heat release rate only change slightly and the two maxima

stay distinct, compared to the increasing broadening of the pressure PDFs. In addition the locations of the maxima of the $\dot{q}'$-PDFs do not change whereas the pressure maxima move to higher amplitudes with increasing noise. These properties indicate saturation in the coupling of unsteady heat release of the flame with the acoustic field.

At the highest equivalence ratio $\phi = 0.715$ (figure 4.14) the pressure PDFs show two very clear maxima at the lowest forcing amplitude. The equivalence ratio is already at the boundary to the bistable region and could be beyond that due to the afore mentioned uncertainty in the setting of ϕ. It seems likely that the PDFs capture noise induced transition to instability (NIT) as it has been shown by Jegadeesan and Sujith (2013). NIT differs from what can be observed here (NIB) at lower equivalence ratios, where triggering is not possible. This difference is illustrated in figure 4.16. Pressure time traces are shown for two different operating conditions, $\phi = 0.708$ (figure 4.16.a,b) and $\phi = 0.715$ (figure 4.16.c,d). The excitation amplitudes amount to $D = 12.7\,Pa$ and $D = 2.8\,Pa$ respectively. Although both signals feature a bimodal PDF the time series reveal that it originates from two different phenomena: At $\phi = 0.715$ the system gets triggered to instability (NIT) eventually and re-stabilises subsequently. The bimodal distribution will only show in the PDF if the transition is captured during the measurement. At $\phi = 0.708$ pressure fluctuates on a smaller time scale throughout the signal; bimodality is only revealed by the PDF. The highlighted PDF in figure 4.16.d shows that the triggered signal almost only features the instability amplitude.

4.3.2 Analyses in the Frequency Domain

Figure 4.17 shows the frequency spectra of the pressure signals for the equivalence ratios $\phi = 0.701$, $\phi = 0.708$ and $\phi = 0.715$ centred around the instability frequency $f_{inst} = 192\,Hz$. Each plot contains three curves for the three noise intensities $D = 5.7\,Pa$, $D = 9.9\,Pa$ and $D = 14.2\,Pa$. The spectra of the non-reactive flow are given for reference. They show an equal increase across all frequencies when the noise intensity grows; the dominant acoustic mode of the duct around $192\,Hz$ is visible.

In the cases of reactive flow a distinct peak at the frequency of the thermoacoustic

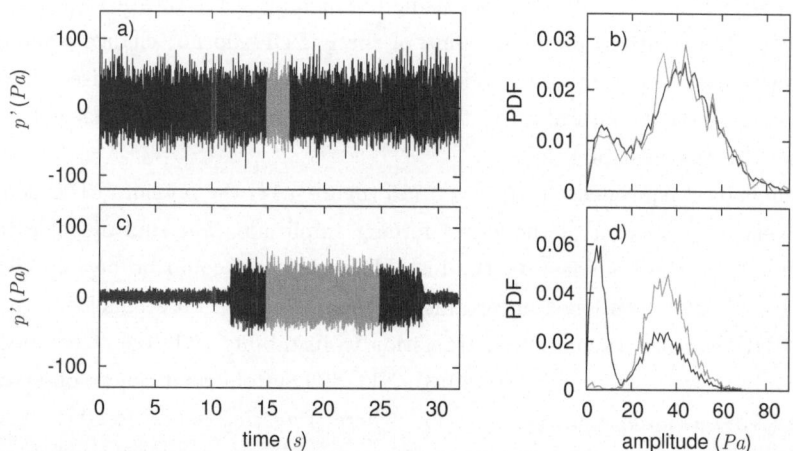

Figure 4.16: Comparison between NIB (a, b; $\phi = 0.708$, $D = 12.7\,Pa$) and NIT (c,d; $\phi = 0.715$, $D = 2.8\,Pa$). PDFs of highlighted areas in a) and c) are represented by grey lines in b) and d) respectively.

instability, $f_{inst} = 192\,Hz$, appears. However a variation of $\delta f_p = \pm\,3\,Hz$ of the peak frequency with respect to changes in the noise intensity can be observed. Around f_{inst} the system response is stronger than at the neighbouring frequencies. For example, at an equivalence ratio of $\phi = 0.708$ the spectrum around $100\,Hz$ only rises from 1 to 10 when the noise intensity increases from $D = 5.7\,Pa$ to $D = 9.9\,Pa$. At the same noise levels the power density at $192\,Hz$, however, increases from 10^2 to about $10^{4.5}$. For a given noise intensity the peak height increases when the equivalence ratio approaches the bistable region.

Wiesenfeld et al. (1995) demonstrated that nonlinear dynamical systems exposed to noise close to a Hopf bifurcation show the frequency spectrum that is expected for the self-excited system beyond the bifurcation. This behaviour is referred to in the literature as a *noisy precursor* of the bifurcation (Neiman et al., 1997; Kiss et al., 2003; Ushakov et al., 2005). The discussed frequency spectra suggest this property to be featured by the investigated combustor.

The ratio of the height H of a peak in the frequency spectrum to its width at half height Δf, normalised by the peak frequency f_p, is called the coherence factor, β (equation 4.1, (Gang et al., 1993; Kiss et al., 2003)). β is used as a

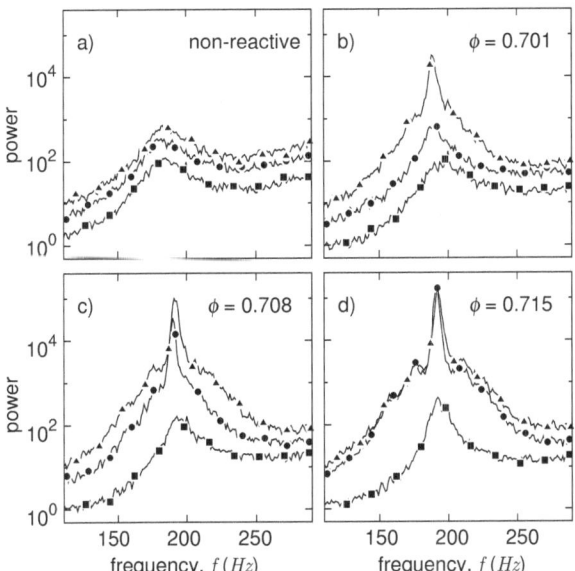

Figure 4.17: Frequency spectra for the system excited by noise at stable conditions close to the bifurcation point (b-c) and for the non-reactive flow (a). Curves indicating three different excitation amplitudes: $D = 5.7\,Pa$ (■), $D = 9.9\,Pa$ (●), $D = 14.2\,Pa$ (▲).

measure for regularity or order in a signal; a pure sinusoidal signal would have a coherence factor of $\beta \rightarrow \infty$, a white noise signal a coherence factor of $\beta = 0$. It is comparable to the signal-to-noise ratio (SNR) used in other publications (Gang et al., 1993; Neiman et al., 1997; Pikovsky and Kurths, 1997).

$$\beta = \frac{H f_p}{\Delta f} \tag{4.1}$$

To derive this measure from the collected data the spectral peak around f_{inst} has been fitted by a Lorentzian line-shape function by means of a least squares approximation (Ushakov et al., 2005); H, Δf, and f_p are then defined by the fitted curve. Figure 4.18 illustrates this method.

Figures 4.19 and 4.20 show the coherence factor β for the acoustic pressure oscillations p' and the heat release rate fluctuations $\dot{q}'$ respectively as a function of the noise intensity D. Each figure contains three curves corresponding to the

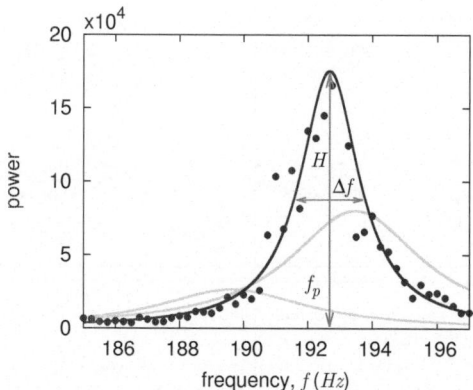

Figure 4.18: Illustration of the Lorentzian fit (solid lines) to the spectral peak around $f = 192\,Hz$.

equivalence ratios $\phi = 0.701$, $\phi = 0.708$, and $\phi = 0.715$.

The two figures show an almost identical picture. For both p' and $\dot{q}'$ the coherence factor of the $192\,Hz$-peak reaches a maximum for an intermediate noise intensity. This can be observed at all three equivalence ratios. However, the optimal noise intensity shifts to lower values as the equivalence ratio increases and the operating conditions approach the bistable region.

This dependency of β on the noise intensity has been observed before for different nonlinear dynamical systems in the context of noise induced phenomena. It was shown first by Pikovsky and Kurths (1997) numerically for the Fitz Hugh–Nagumo system and has been referred to since as *coherence resonance*. Neiman et al. (1997) showed it numerically for a white noise driven Rössler system close to a period-doubling bifurcation. It was also shown experimentally by Kiss et al. (2003) for an electrochemical system and by Zakharova et al. (2013) for an electronical system.

By showing this behaviour to be present in the studied system, underlines that the generality of coherence resonance extends to the field of thermoacoustic systems.

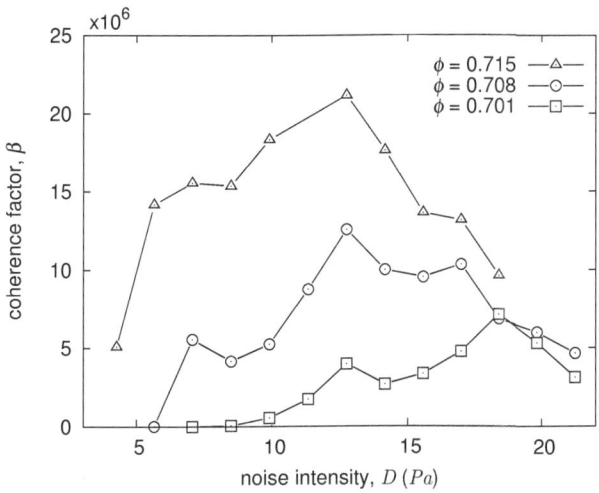

Figure 4.19: Coherence factor β for the acoustic pressure p' as a function of the noise intensity D.

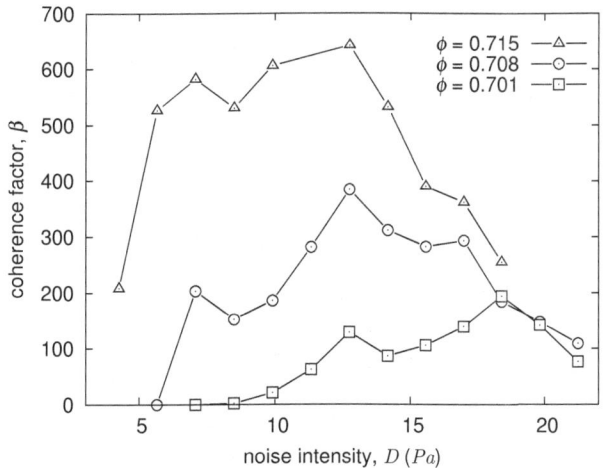

Figure 4.20: Coherence factor β for the heat release rate fluctuations $\dot{q}'$ as a function of the noise intensity D.

5 Conclusion

Combustion systems such as gas turbines, rocket motors, and industrial furnaces are known for their ability to develop a self-sustaining feedback loop between the unsteady heat release of the combustion process and the pressure fluctuations of the surrounding flow field. This phenomenon, called thermoacoustic instability, produces pressure fluctuations of large amplitudes which are bound by nonlinear effects that are hard to predict during the design phase of a combustor. Thermoacoustic instability is detrimental to a safe and reliable operation of the respective application.

The operating conditions which lead to the onset of an instability define the stability boundary of the system. This boundary restricts the application in its feasible range of operating conditions and thus knowledge of its properties is crucial to improve industrial combustors.

This work addresses the effect of acoustic white noise on a thermoacoustic system close to its stability boundary. To this purpose an existing methane fuelled, premixed model combustor featuring a laminar, quasi-flat flame at the energy lab of ISTA, TU Berlin, has been modified to operate with a minimum of intrinsic noise.

By variation of the fuel mass flow rate, and thus the equivalence ratio, the stability characteristics of the unforced system under lean conditions have been determined. It has been found that the system enters thermoacoustic instability through a subcritical Hopf bifurcation, and leaves that state through a saddle-node bifurcation. These two bifurcations define three regions of qualitatively different stability: For values of the equivalence ratio below the saddle-node bifurcation the system is *stable* and remains in an unexcited state; once the excited system enters this region it will evolve exponentially towards the unexcited state. The region between the saddle-node and the Hopf bifurcation is considered linearly stable: It remains in the unexcited state for small acoustic disturbances. Pulses of larger amplitude, however, will lead to the system leaving the unexcited state end entering the self-sustained limit cycle oscillation characteristic to the thermoacoustic instability. This phenomenon call triggering or hard excitation is known for systems that undergo a subcritical Hopf bifurcation. Due to these two states the system can potentially exist in, this domain of the equivalence ratio is denominated *bistable* region.

The region of equivalence ratios equal to and above the Hopf bifurcation point is considered *unstable*: Disturbances to the stable system of any amplitude will result in the system leaving that state and exhibiting the thermoacoustic instability.

Acoustic white noise has been imposed on the flow field upstream of the flame at equivalence ratios in the stable region and at the border to the bistable region. It has been observed that the system responds to the random excitation with pressure and heat release rate oscillations at the same frequency as during thermoacoustic instability, $192\,Hz$. Both variables show the same qualitative dependency of their oscillation magnitude (quantified by the rms value of 32 second time series) on the equivalence ratio and noise intensity: Above a certain noise level they increase monotonically towards the magnitude observed during instability, exceeding the excitation amplitude.

Probability density functions of the amplitude distributions of acoustic pressure and heat release rate time series revealed two maxima manifesting for both variables when the stable system was excited with noise. This noise induced bimodality has been observed before numerically and experimentally for the noise driven Duffing-Van der Pol oscillator close to its stability boundary. However, it has not been shown before for a thermoacoustic system.

Investigating the frequency domain of the data showed that the system responds to the stochastic noisy excitation with coherent motions, namely pressure and heat release rate fluctuations with a prevailing frequency of $192\,Hz$, which is the dominant frequency of the thermoacoustic instability. This similarity between the noise induced spectrum of the stable system and the spectrum of the unstable system in the absence of noise has been referred to as *noisy precursora* to the bifurcation in the literature.

Further, it could be shown that the coherence factor, quantifying the degree of order in the pressure and heat release rate fluctuations, shows a maximum for intermediate noise intensities. This optimal excitation intensity reduces when the equivalence ratio approaches the bistable region. This observation at the investigated system correlates to findings in other nonlinear, excitable systems driven by noise. It has been referred to as *coherence resonance*.

The results obtained in this work support the generality of noise induced phenomena in the domain of thermoacoustic systems. The gathered, analysed, and discussed data aids in the understanding of fundamental dependencies of thermoacoustic systems on noisy acoustic excitations. Further research could address intrinsic noise specific to combustors of larger scale (coloured noise) and its impact on systems closer to industrial applications. Establishing a theoretical model that captures the effects of additive noise on thermoacoustic systems is another topic that could benefit from this work.

Bibliography

Arnold, L.
2013. *Random dynamical systems*. Springer Science & Business Media.

Barrere, M. and F. Williams
1969. Comparison of combustion instabilities found in various types of combustion chambers. In *Symposium (International) on Combustion*, volume 12, Pp. 169–181. Elsevier.

Benzi, R., G. Parisi, A. Sutera, and A. Vulpiani
1982. Stochastic resonance in climatic change. *Tellus*, 34(1):10–16.

Benzi, R., A. Sutera, and A. Vulpiani
1981. The mechanism of stochastic resonance. *Journal of Physics A: mathematical and general*, 14(11):L453.

Burnley, V. S. and F. E. C. Culick
2000. Influence of random excitations on acoustic instabilities in combustion chambers. *AIAA Journal*, 38(8):1403–1410.

Chiu, H., E. Plett, and M. Summerfield
1973. Noise generation by ducted combustion systems. In *Aeroacoustics Conference*. American Institute of Aeronautics and Astronautics.

Chiu, H. and M. Summerfield
1974. Theory of combustion noise. *Acta Astronautica*, 1:967 – 984.

Douglass, J. K., L. Wilkens, E. Pantazelou, F. Moss, et al.
1993. Noise enhancement of information transfer in crayfish mechanoreceptors by stochastic resonance. *Nature*, 365(6444):337–340.

Dowling, A. P.
1997. Nonlinear self-excited oscillations of a ducted flame. *Journal of Fluid Mechanics*, 346:271–290.

Dowling, A. P.
1999. A kinematic model of a ducted flame. *Journal of fluid mechanics*, 394:51–72.

Fauve, S. and F. Heslot
1983. Stochastic resonance in a bistable system. *Physics Letters A*, 97(1):5–7.

Gang, H., T. Ditzinger, C. Ning, and H. Haken
1993. Stochastic resonance without external periodic force. *Physical Review*

Letters, 71(6):807.

Haber, L. C.
2000. An investigation into the origin, measurement and application of chemi-luminescent light emissions from premixed flames.

Horsthemke, W.
1984. *Noise induced transitions*. Springer.

Huang, Y. and V. Yang
2009. Dynamics and stability of lean-premixed swirl-stabilized combustion. *Progress in Energy and Combustion Science*, 35(4):293 – 364.

IEA
2014. *Key World Energy Statistics*. IEA.

Jegadeesan, V. and R. Sujith
2013. Experimental investigation of noise induced triggering in thermoacoustic systems. *Proceedings of the Combustion Institute*, 34(2):3175 – 3183.

Juniper, M. P.
2010. Triggering in the horizontal rijke tube: non-normality, transient growth and bypass transition. *Journal of Fluid Mechanics*, 667:272–308.

Kabiraj, L., A. Saurabh, R. Steinert, and C. O. Paschereit
2013. Instabilities in a confined flat-flame system. *Proceedings of the European Combustion Meeting*.

Kabiraj, L. and R. Sujith
2011. Investigation of subcritical instability in ducted premixed flames. In *ASME 2011 Turbo Expo: Turbine Technical Conference and Exposition*, Pp. 969–977. American Society of Mechanical Engineers.

Kabiraj, L. and R. I. Sujith
2012. Nonlinear self-excited thermoacoustic oscillations: intermittency and flame blowout. *Journal of Fluid Mechanics*, 713:376–397.

Kiss, I. Z., J. L. Hudson, G. J. E. Santos, and P. Parmananda
2003. Experiments on coherence resonance: Noisy precursors to hopf bifurcations. *Physical Review E*, 67(3).

Lieuwen, T. C.
1999. *Investigation of combustion instability mechanisms in premixed gas turbines*. PhD thesis, Georgia Institute of Technology.

Lieuwen, T. C.
2002. Experimental investigation of limit-cycle oscillations in an unstable gas turbine combustor. *Journal of Propulsion and Power*, 18(1):61–67.

Lieuwen, T. C. and A. Banaszuk
2005. Background noise effects on combustor stability. *Journal of Propulsion and Power*, 21(1):25–31.

Lieuwen, T. C. and V. Yang, eds.
2005. *Combustion instabilities in gas turbine engines(operational experience, fundamental mechanisms and modeling)*, volume 210. American Institute of Aeronautics and Astronautics.

Matveev, K.
2003. *Thermoacoustic instabilities in the Rijke tube: experiments and modeling.* PhD thesis, California Institute of Technology.

McNamara, B., K. Wiesenfeld, and R. Roy
1988. Observation of stochastic resonance in a ring laser. *Physical Review Letters*, 60(25):2626.

Merk, H.
1957. An analysis of unstable combustion of premixed gases. In *Symposium (International) on Combustion*, volume 6, Pp. 500–512. Elsevier.

Moeck, J., M. Oevermann, R. Klein, C. Paschereit, and H. Schmidt
2009. A two-way coupling for modeling thermoacoustic instabilities in a flat flame rijke tube. *Proceedings of the Combustion Institute*, 32(1):1199–1207.

Moeck, J. P., M. R. Bothien, S. Schimek, A. Lacarelle, and C. O. Paschereit
2008. Subcritical thermoacoustic instabilities in a premixed combustor. *AIAA paper*, 2946.

Neiman, A., P. I. Saparin, and L. Stone
1997. Coherence resonance at noisy precursors of bifurcations in nonlinear dynamical systems. *Physical Review E*, 56(1):270.

Noiray, N., M. Bothien, and B. Schuermans
2011. Investigation of nonlinear mechanisms driving combustion instabilities using an electro-acoustic van der pol oscillator. In *7th European Nonlinear Oscillations Conference (ENOC 2011), Rome, July*, Pp. 24–29.

Noiray, N. and B. Schuermans
2012. Deterministic quantities characterizing noise driven hopf bifurcations in gas turbine combustors. *International Journal of Non-Linear Mechanics*, 50(0):152 – 163.

Pikovsky, A. S. and J. Kurths
1997. Coherence resonance in a noise-driven excitable system. *Physical Review Letters*, 78(5):775.

Rayleigh, J. W. S. B.
1896. *The theory of sound*, volume 2. Macmillan.

Rijke, P.
1859. Notiz über eine neue art, die in einer an beiden enden offenen röhre enthaltene luft in schwingungen zu versetzen. *Annalen der Physik*, 183(6):339–343.

Strahle, W. C.
1978. Combustion noise. *Progress in Energy and Combustion Science*, 4(3):157 – 176.

Straub, D. L. and G. A. Richards
1998. Effect of fuel nozzle configuration on premix combustion dynamics. In *ASME 1998 International Gas Turbine and Aeroengine Congress and Exhibition*, Pp. V003T06A044–V003T06A044. American Society of Mechanical Engineers.

Strogatz, S. H.
2000. *Nonlinear Dynamics and Chaos: With Applications to Physics, Biology, Chemistry and Engineering*. Westview Press.

Subramanian, P., R. I. Sujith, and P. Wahi
2013. Subcritical bifurcation and bistability in thermoacoustic systems. *Journal of Fluid Mechanics*, 715:210–238.

Tacina, R.
1990. Low no(x) potential of gas turbine engines. In *Aerospace Sciences Meetings*, Pp. –. American Institute of Aeronautics and Astronautics.

Ushakov, O. V., H.-J. Wünsche, F. Henneberger, I. A. Khovanov, L. Schimansky-Geier, and M. A. Zaks
2005. Coherence resonance near a hopf bifurcation. *Phys. Rev. Lett.*, 95:123903.

Wellens, T., V. Shatokhin, and A. Buchleitner
2004. Stochastic resonance. *Reports on Progress in Physics*, 67(1):45.

Wicker, J. M., W. D. Greene, S.-I. Kim, and V. Yang
1996. Triggering of longitudinal combustion instabilities in rocket motors-nonlinear combustion response. *Journal of Propulsion and Power*, 12(6):1148–1158.

Wiesenfeld, K., F. Moss, et al.
1995. Stochastic resonance and the benefits of noise: from ice ages to crayfish and squids. *Nature*, 373(6509):33–36.

Zakharova, A., A. Feoktistov, T. Vadivasova, and E. Schöll
2013. Coherence resonance and stochastic synchronization in a nonlinear circuit near a subcritical hopf bifurcation. *The European Physical Journal Special Topics*, 222(10):2481–2495.

Zakharova, A., T. Vadivasova, V. Anishchenko, A. Koseska, and J. Kurths
2010. Stochastic bifurcations and coherencelike resonance in a self-sustained bistable noisy oscillator. *Phys. Rev. E*, 81:011106.

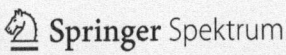